VICTORIAN AND EDWARDIAN LOCOMOTIVE PORTRAITS
NORTHERN ENGLAND, WALES, SCOTLAND AND IRELAND

VICTORIAN AND EDWARDIAN LOCOMOTIVE PORTRAITS

NORTHERN ENGLAND, WALES, SCOTLAND AND IRELAND

Anthony Burton

PEN & SWORD TRANSPORT

AN IMPRINT OF PEN & SWORD BOOKS LTD.
YORKSHIRE – PHILADELPHIA

First published in Great Britain in 2024 by
Pen and Sword Transport
An imprint of
Pen & Sword Books Ltd.
Yorkshire - Philadelphia

Copyright © Anthony Burton, 2024

ISBN 978 1 03610 051 3

The right of Anthony Burton to be identified as author of this work has been asserted by him in accordance with the Copyright, Designs and Patents Act 1988.

A CIP catalogue record for this book is available from the British Library.

All rights reserved. No part of this book may be reproduced or transmitted in any form or by any means, electronic or mechanical including photocopying, recording or by any information storage and retrieval system, without permission from the Publisher in writing.

Typeset by SJmagic DESIGN SERVICES, India.
Printed and bound in India by Parksons Graphics Pvt. Ltd.

Pen & Sword Books Ltd. incorporates the imprints of Pen & Sword Books: After the Battle, Archaeology, Atlas, Aviation, Battleground, Discovery, Family History, History, Maritime, Military, Naval, Politics, Railways, Select, Transport, True Crime, Fiction, Frontline Books, Leo Cooper, Praetorian Press, Seaforth Publishing, Wharncliffe and White Owl.

For a complete list of Pen & Sword titles please contact

PEN & SWORD BOOKS LIMITED
George House, Units 12 & 13, Beevor Street, Off Pontefract Road,
Barnsley, South Yorkshire, S71 1HN, England
E-mail: enquiries@pen-and-sword.co.uk
Website: www.pen-and-sword.co.uk

or

PEN AND SWORD BOOKS
1950 Lawrence Rd, Havertown, PA 19083, USA
E-mail: uspen-and-sword@casematepublishers.com
Website: www.penandswordbooks.com

Contents

Acknowledgements	6
Introduction	7

Photo Section

Wales	14
Secondary Joint Lines	33
North of England	40
Isle of Man	56
British Light Railways	58
Scotland	78
Irish Railways	97
Index	111

Acknowledgements

The photographs used in this book are from the John Scott Morgan Collection apart from the following: Hopwood Collection, pp. 27, 49; Photomatic, pp. 28, 32, 68; RAS Collection. pp. 54, 78-81, 88; Rev. E.R. Boston Collection, p. 65.

Introduction

The railway age could be said to have its origins in the horse-drawn tramways of the North of England and Wales. It was on the Penydarren Tramway in South Wales that Richard Trevithick gave the first public demonstration of his steam locomotive. Unfortunately, the engine broke the brittle cast iron rails, and development had to wait a few more years. The Napoleonic Wars caused rocketing prices for fodder for the horses, so the owner of the Middleton Colliery near Leeds decided he would try to use locomotives on his line that linked the mine to the Aire & Calder Navigation. He paid for the Trevithick patent but still had to overcome the problem of breaking rails. His answer, arrived at in collaboration with the engineer Matthew Murray, was to use a rack and pinion system, where a cog on the engine meshed with a toothed rail. This gave enough traction for even a very light engine to work. The rack and pinion next appeared in 1896 with the opening of the Snowdon Mountain Railway. When the first steam locomotives went into service in August 1812, they attracted huge interest. Among those who went to see the engines and take notes was George Stephenson of the Killingworth Colliery.

Stephenson became one of a small number of engineers who decided to continue developing locomotives for use on the tramway system, but it was Stephenson who persevered. His success culminated with his work as chief engineer for the first public railway that was authorised by Act of Parliament to use steam locomotives, the Stockton & Darlington. When it opened in 1825, locomotive use was still restricted to freight – passengers were carried in a conventional horse-drawn stage coach, fitted with flanged wheels. This was still, in effect, merely another colliery line and there had been very little real development, apart from the abandonment of the rack and pinion. The first locomotives, *Locomotion* and *Hope* had single flue boilers, in which the firebox was at one end and the end of the flue was turned up vertically to create the chimney. The engine was carried on four coupled wheels in an 0-4-0 configuration. There were two vertical cylinders set in line, with cross beams above them. From these, connecting rods led down to crank pins on the wheels, set at quarter turns from each other. There was no actual reversing gear; the driver had to disconnect the valve motion by hand and reset it to go in the opposite direction. At the opening ceremony a commentator recorded that the speed of the engine, with its train of guests travelling in adapted coal trucks, was 12mph, but that was on the downhill run to Darlington. What was shown that day was that a locomotive could haul a train full of passengers – estimations of numbers varied from 450 to six hundred. In working practice, however, overall speeds were slower and the engines were always likely to run out of steam. They did well enough for what was still basically a colliery line, but for railways to develop into major transport routes, something much better would be needed.

When proposals were being discussed for what would be a very different line to link Manchester and Liverpool, there was a great debate on how it should be run. One faction was unconvinced by current locomotive performance and favoured a system by which trains would be hauled along by cable worked by a series of stationary engines dotted along the track. Experts failed to agree on which was the better system, so it was decided to hold a locomotive trial on part of the track at Rainhill. It was obvious that if locomotives were to win the contest, they would have to perform a great deal better than those already in use, So, conditions were laid down. Engines carried on six wheels could weigh up to 6 tons, if on four wheels, 4½ tons. The 6 ton engines had to draw a train of gross weight 20 tons at 10mph over a distance equivalent to that between Liverpool and Manchester. Boiler pressure must not exceed 50 pounds per square inch (p.s.i.) and the safety valve had to be out of reach of the driver.

In the end, there were only three contenders for the trial: a light weight, vertically boiled engine, *Novelty*; *Sans Pareil* a conventional engine for the time, that mainly differed from the earlier engines in that there was a return flue, bent round in a U-shape, so that the firebox was at the chimney end; and the Robert Stephenson designed *Rocket*.

As is well known, *Rocket* was the winner, comfortably meeting all the requirements and at one time gave a dramatic demonstration of its power by speeding down the track at an unprecedented 30mph. The importance of *Rocket*'s win was not simply that it proved locomotives could do the work demanded of them, but that the elements that made it possible were all vital for later development. It was obvious that efficient steaming was the essential for creating greater power. The return flue boiler had been the most efficient available, but the new engine had a multi-tubular boiler, in which instead of a single flue, a large number of small tubes carried the hot fumes from the firebox through the water, greatly increasing the surface area of hot metal. To increase the efficiency, the exhaust steam from the pistons, instead of simply escaping into the open air, was led through a blast pipe to the base of the chimney, drawing more air through the firebox. Unlike *Locomotion*, it had a reversing gear, operated by a foot pedal on the footplate – but no brakes. The pistons were inclined at 45° driving directly down to cranks on the wheels. They would later be lowered to nearer the horizontal.

The locomotives described in the following pages all have their origins in these early developments and in particular in the innovations introduced in *Rocket* – the multi tubular boiler, exhaust pipe blast and angled or horizontal cylinders driving the wheels through connecting rods from the piston. There would be many improvements, notably in valve gear. The Stephenson gear developed in 1842 was not just designed to allow the engine to be reversed. It allowed for variable cut off of the steam entering the cylinder. In full forward, the gear allowed steam to enter the cylinder for the full length of the piston stroke, valuable for times such as starting the engine moving. Once under away, however, the steam could be cut off part way down the stroke, allowing the natural expansion to work for the rest of the piston's travel, which was far more efficient in terms of fuel economy. A useful analogy is in driving a car that starts in first gear and then moves to higher gears when under way.

It was not just improvements in locomotive design that marked the changes begun in the early years that were to have a long-term effect on railway travel. Trains need good track to run on. The early tramways had rails carried on stone blocks with central holes that were filled with wooden plugs, into which the rails could be spiked. This allowed for a free space between the rails for the horses to walk. The same system was continued on the early railways for a time. The tramways were all self-contained units, mostly linking a colliery or works to a navigable waterway. There was no need to consider what gauge was being used elsewhere. So, when George Stephenson built his first locomotive, it ran on the Killingworth colliery line that had a 4ft 8in gauge. When he came to work on the Stockton & Darlington, rather than sit down to consider what might be the best gauge for a new line, he simply kept to the same measurement, though an extra half inch crept in somehow. There was, however, to be a major change in the rails themselves, with the introduction of wrought iron to replace the brittle cast iron of the first track. The problem of rail cracking was reduced, allowing the construction of heavier, more powerful engines. In time they would be replaced by even more durable steel.

In time, the stone sleeper blocks gave way to wooden transverse sleepers on northern railways. Isambard Brunel famously favoured not only a different gauge, but also a different way of mounting sleepers. His broad 7ft gauge track was mounted on longitudinal wooden sleepers, running the full length of the track. As long as his Great Western Railway had no junctions with other lines, this was not a problem. But where they met, everything, passengers as well as goods, had to be transferred from one system to the other. This was such an inconvenience that one or other had to go. This led to competition as neither side wanted to be the ones that were scrapped and that proved a spur to locomotive development, each side wanting to demonstrate that their services were smoother and faster than their rival's. There was a trial, but as it was incapable of showing whether better performance was due to more efficient locomotives or better track, it proved nothing. The end was, in any case, inevitable. The cost of converting the Stephenson gauge to broad would have been ruinously expensive, whereas all that was needed to allow all trains to run on the broad gauge was to lay a third rail between the existing two. That was a temporary measure before broad gauge vanished forever from mainland Britain in 1892, apart from a short reconstructed section at the railway museum at Didcot.

The Battle of the Gauges as it came to be known was not the only competition that led to keen rivalries between different companies. Mostly, railway companies were content to work as they pleased without worrying too much about what others were doing. After all, if passengers wanted to travel from, say, York to Manchester, they would not change onto a train heading from York to Newcastle if it happened to be offering a faster service. But when it came to routes from London to Scotland, matters were very different. Here passengers had a choice of routes, leaving King's Cross for the east coast line or Euston to travel up the west coast. This resulted in what became known as the 'Races to the North'. In July and August 1888, companies began running light express trains. The Great Northern ran the fastest time up the east coast covering the 392 miles to Edinburgh in 17 hours 27 minutes, an average speed of over 60mph. The London and North Western Railway (LNWR) on the west coast had a slightly longer journey of 400 miles and had a marginally slower time. On 22 August 1895, there was a race from London to Aberdeen and this time the honours went to the LNWR on the west coast, with a time of 18 hours 23 minutes, with the Great Northern just losing out by a quarter of an hour. On the LNWR, there was a heroic effort by the locomotive *Hardwicke* that took the section from Crewe to Carlisle, that included climbing over the notoriously steep section to the Shap Fell Summit, but still achieved an average speed of 67.2mph. Among the engines flying the Great Northern colours were the famous Singles designed by Patrick Sterling, not merely powerful engines but arguably the most elegant of all Victorian steam locomotives. Whatever else it achieved, the competition resulted in some outstanding designs.

Although the Stephenson gauge was accepted as the standard for mainland Britain before the end of the nineteenth century, it was not the only gauge in use. In the tramway age, there had been a variety of gauges in use and in Wales the Ffestiniog Railway, linking the slate mines of Blaenau Ffestiniog to the harbour at Porthmadog was built to the 1ft 11½in gauge. It climbed the hillside from sea level to a height of 800ft and the track contained several severe curves. This was not a problem in the days when it was worked by horses. The slate traffic was all downhill, so could descend by gravity and the horses were only required to haul the empty waggons back to the top. There was a problem, however, when it came to changing from horses to locomotives. Only short wheel-based engines could negotiate the curves. Once more power was needed, as the line also began to carry passengers, a new form of locomotive was required – the double-ended Fairlies, which can be seen on pages 22 and 23. So, mainland Britain had, for a time, three gauges in use – broad, standard and narrow.

When it came to railway building in Ireland, there was no need to accept any of the gauges then in use on the other side of the Irish Sea. Engineers could decide on what they considered the best option and build a coherent system. The Dublin & Kingston Railway was initially built to the British standard gauge, even though Stephenson himself, when asked what would be the ideal gauge for Ireland, had suggested somewhere between 5ft and 5ft 6in. The Dublin & Drogheda took that advice and opted for 5ft 2in, while the Ulster Railway had decided on a more imposing 6ft 2in. This was clearly unsatisfactory, and the Board of Trade stepped in. They decided on an average of the three lines then being built and the new Irish standard gauge was set at 5ft 3in. As in the rest of

Britain, a number of independent narrow gauge lines were also constructed.

There was to be one other major change in the way in which railways were constructed that had an effect on the locomotives that were to use the line. At the end of the nineteenth century, Britain had an extensive rail network, but there were still areas, mainly rural, that were still not served. It did not seem worthwhile for companies to go the expense of obtaining an Act of Parliament and building lines that met the exacting standards required by tracks accepting heavy goods trains and expresses. All that changed with the passing of the Light Railway Act of 1896. Companies were allowed to build lines without the formalities of an Act to be debated in parliament. They could use lighter rails and did not need to conform to the standard gauge but could suit whatever seemed appropriate for the particular district. Many other economies could also be made. There was, however, a downside to this. Speeds were limited and passengers were provided with carriages that offered a good deal less comfort than available on most other trains, but then journeys were mainly quite short. As a result, over the years, lines were built to different gauges and served by a motley array of locomotives as can be seen in the illustrations between pages 58 and 77.

The range of locomotives built in these different regions was immense. Partly it was due to the way in which track and gauges changed, partly due to the inevitable progress in new developments. It was also due, in good measure, to the multiplicity of companies, each employing their own chief mechanical engineers with their own ideas on design.

This, of course, makes them all the more interesting. Many new ideas were tried out over the years, some of which proved quite controversial. One on which engineers rarely agreed was on the value of compounding. This was not a new concept. It had first been developed in stationary steam engines before even the first locomotive had been built. Locomotive development depended on using steam at high pressure – and that pressure increased steadily over the years. But even at the comparatively low pressure of 50psi, stipulated for engines at the Rainhill Trials, the steam was still under pressure when it was exhausted from the cylinder. In compound engines, that steam, instead of being immediately passed down the blast pipe to the base of the chimney, was fed instead to a second low pressure cylinder. Henry Ivatt, for example, developed the Great Northern C1 Class in 1898 as conventional 4-4-2 locomotives, but in 1903 adapted them as 4-cylinder compounds, with two high pressure and two low-pressure cylinders. Only three of the compounds were built and when Ivatt developed his next 4-4-2 locomotives, the C2s, he reverted to two cylinders and a total of sixty were built.

Other innovations that improved performances were superheaters and air brakes. Steam in a conventional boiler is saturated, that is, it contains water droplets. By passing it though a superheater before it reaches the cylinder, the water is entirely removed and the result could be an increase of efficiency of as much as 25 per cent. The idea was first discussed in the middle of the nineteenth century, but it was only in the Edwardian period that a really efficient superheater was developed. The air brake that used compressed air to activate the brakes was patented by

George Westinghouse in 1869 and soon came into general use, making high speed travel considerably safer.

Although the major companies had their own locomotive works and looked to their engineers for improvements and developments, smaller companies usually relied on buying engines from specialist manufacturers, a process that began on a large scale when the Robert Stephenson works were established in Newcastle in 1823. So, throughout the following pages, the names of many different companies will appear as manufacturers alongside those of the railway companies themselves. All these different factors combine to provide a rich mixture and great variety of locomotives, ranging from humble tank engines intended for light duties to the glamorous express locomotives and the less glamorous but no less essential freight locomotives intended for heavy duty.

PHOTO SECTION

Wales

Taff Vale Railway 0-6-0 Tender Goods No.8. The railways of South Wales are normally associated with busy tank engine designs. However, a number including the Taff Vale and the Rhymney Railway had tender engines, with a small number lasting long enough to be absorbed into a greater Great Western at the grouping in 1923.

Rhymney Railway Class AP 0-6-2 tank No.36 was designed by Charles Hurry Riches and most of the class were constructed by Hudswell Clark of Leeds, from 1911 onwards, with some later modifications. These locomotives were used on passenger and heavy coal trains in the South Wales valleys. It became GWR (Great Western Railway) No.79 in 1923 and was withdrawn by the British Railways, Western Region in August 1955.

Hudswell Clark constructed steam rail motor, No.1 built in 1907 for use on the Aber Junction to Sendghenydd branch. The locomotive portion was later reused as a small shunting tank locomotive when the steam rail motors were withdrawn.

Brecon & Merthyr Junction Railway 0-6-2 tank No.43 was designed by James Dunbar and constructed by Robert Stephenson of Newcastle in April 1914. No.43 is seen at Brecon Shed's turntable. Absorbed into the Great Western in 1923, members of the class were 'Swindonised' by being fitted with Swindon Standard Number 2 boilers. A simple and useful design, the Collett 56XX 0-6-2 tank would draw much of its inspiration from the Birmingham & Midland Joint Railway's small fleet of capable 0-6-2 tank designs designed by the Rhymney Railway's Chief Engineer Hurry Riches and the Brecon's own James Dunbar. Under the GWR the 0-6-2 became No.1113 and later No.426. The engine was finally withdrawn on the eve of nationalisation on 31 October 1947.

The Barry Railway was incorporated in 1884 to provide an alternative line to the Taff Valley Railway. At its full extent it had a total length of 70 route miles. One of the key promoters was David Davies, who was the main contractor and promoter of the Cambrian Railway and was also the owner of several collieries including the famous Ocean mine. The railway had junctions with the Taff Vale, Rhymey and Brecon & Merthyr Railways. The most important feature was the Barry Docks, a deep-water facility able to take large ships. In 1913 it had a record-breaking export of 11 million tons of coal. It became part of the GWR at the railway grouping of 1923. Locomotive No.89 was a J Class 2-4-2 tank designed by J.H. Hosgood and constructed 1898 by Sharp Stuart. The J Class was the last passenger type to be built for the Barry Railway, being used on local services. It was renumbered as 1913 by the GWR and withdrawn from service in October 1896.

Cambrian Railways, Sharp Stuart of Manchester, later Glasgow, constructed 0-6-0 tender goods *Marquis,* named after the son of the Fifth Marquess of Londonderry, the Chairman of the Company. The 0-6-0 is depicted in original condition. Copying broad gauge GWR practice, Cambrian locomotives at first received only names, not numbers. The Cambrian Railway purchased mostly 0-6-0 tender goods, 2-4-0 tanks and some 4-4-0 tender locomotives principally from Sharp Stuart. Most were absorbed into the GWR in 1923, though *Marquis* was not so fortunate. These distinctive locomotives were used on both goods and passenger work.

0-4-0 box tank No.1 *The Princess* was constructed by George England at its Hatcham Iron Works, New Cross London. No.1 is seen shortly after delivery to the Festiniog Railway in 1863. In later years the railway changed its name to the Welsh version and became the Ffestiniog.

Princess as reconstructed as a saddle tank, but still retaining its open cab and other George England features, notably its original style of chimney. It is seen here at Blaenau Ffestiniog.

George England constructed Double Fairlie 0-4-0 0-4-0 articulated locomotive as built in 1869, and scrapped in 1882. This was the first of the famous double Fairlie locomotive on the Festiniog and is here seen at Port Madoc, Harbour Station with a passenger train. c1869.

A James Spooner designed double Fairlie locomotive as constructed by the Avonside Locomotive Company in 1872. The last Fairlie locomotive on the Festiniog Railway to be rebuilt with a wagon top boiler in 1908, the locomotive was finally withdrawn in 1930. The engine is here pictured outside the Company's Boston Lodge Works c.1890.

Single Fairlie 0-4-4 tank locomotive *Taliesin* constructed in 1876 by Vulcan foundry, Newton Le Willows. It was withdrawn in 1924 and is here seen at Portmadoc Harbour Station c.1890.

Ffestiniog and Blaenau Railway Manning Wardle constructed 0-4-2 saddle tank No.2. The 0-4-2 is here seen on a passenger train of four wheeled carriage stock. The narrow gauge line opened in 1868 and connected with the more famous Festiniog 1ft 11½in narrow gauge at Blaenau. The railway had two identical Manning Wardle steam locomotives named *Nipper* and *Scorcher*. Never as busy as the Festiniog proper, the 1ft 11¼in narrow railway was purchased by the Great Western Railway in 1876, who then promptly converted the line to 4ft 8½in standard gauge, after which the original locomotives and stock were scrapped.

Fletcher Jennings 0-4-2 saddle tank, Talyllyn Railway No.1 *Talyllyn* is here seen at Towyn Wharf Station c.1895. The Talyllyn Railway was opened in 1865 and still has a pair of Fletcher Jennings tank locomotives in service! Constructed at Whitehaven in 1865, No.1 was a 0-4-2 saddle tank and No.2 an ungainly 0-4-0 well tank.

Talyllyn Railway No.2 *Dolgoch* at Abergynolwyn c1910. This displays the rather ungainly look of this 1865 locomotive to good effect. This was the only locomotive in working order when the preservation society first took over the line in 1951. However, it was subsequently discovered that although the boiler inspector had passed No.2, he had test bored into one of the only sound pieces of metal remaining in the boiler barrel and that in reality it should have been condemned on the spot.

The Corris Railway of North Wales had a fleet of three Hughes Falcon designed 0-4-2 saddle tanks. These had originally been supplied as 0-4-0 saddle tank locomotives with colonial style open cabs in 1878. An unidentified member of the trio is seen at the original Machynlleth terminus of the line, with the station being later reconstructed in 1908.

The Welshpool & Llanfair Light Railway was opened in 1908 to serve the agricultural area centred around the important Mid-Wales market town of Welshpool. The railway was operated by the Cambrian Railways on opening and featured two Beyer Peacock 0-6-0 tanks, No.1 *The Earl* and No.2 *The Countess* constructed at their Gorton Works. The line was taken over by the Great Western in 1923 and the locomotives were renumbered as 822 and 823. Under British Railways Western Region, the line closed on 3 November 1956, though the last passenger services had ended on 9 February 1931.

Vale of Rheidol Light Railway Davis and Metcalf 2-6-2 tank *Edward VII* constructed in 1900 and seen at the original terminus of the line in Aberystwyth c.1901. The Vale of Rheidol Light Railway was designed to serve the lead and tourist industry in the Rheidol valley. It was later taken over by the Cambrian Railways on 1 July 1913 and by the Great Western at the grouping in 1923. The line became part of British Railways Western Region in 1948 and later the London Midland Region in 1963, before being privatised in 1989. One of a pair, the other being named *Prince of Wales*, they were absorbed into the GWR fleet in 1923 and renumbered as No.1212 and No.1213. The railway had three locomotives, two Davis & Metcalf 2-6-2 tanks and one small 2-4-0 tank named *Rheidol* which was supplied by W.G. Bagnall of Stafford which was second hand from the Plynlimon & Hafan Tramway. *Edward VII* was scrapped in the mid-1920s and was replaced by two new Great Western 2-6-2 tanks and a rebuilt No.1213, later No.9 *Prince of Wales*.

Snowdon Mountain Railway 0-4-2 rack tank No.2 *Enid* at the lower terminus c.1900. Note the distinctive slanting boiler set at 9 degrees to ensure the crown of the firebox remains covered at all times during the climb and decent from the mountain side. The original locomotive fleet was all supplied from Switzerland during 1895/6 for the opening on 6 April 1896. Locomotive No.1 *Ladas* was destroyed in an accident on opening day when it derailed and crashed down the mountain with the loss of one life. The line ran from Llanberis to the summit of Mount Snowdon and is still in operation using a mixture of steam, diesel and battery traction.

The Glyn Valley Tramway was a road side tramway serving local slate quarries and the agricultural community in the rural Glyn Valley in Denbighshire, north east Wales. The tramway had a fleet of Beyer Peacock constructed 0-4-2 tram locomotives which operated the passenger and the goods operations on a line with the unusual gauge of 2ft 4½in. The Dennis family who owned the tramway were also involved with the narrow gauge Snailbeach District Railway in Shropshire. One of the three tram locomotives used on the line c.1895, *Dennis*, *Glyn* and *Sir Theodore* still sported their original attractive lined green livery. Note the skirting covering over the wheels, the sign of a true tram locomotive. The line acquired an ex-War Department American Baldwin 4-6-0 tank after the First World War. The tramway eventually closed in July 1935.

Secondary Joint Lines

The Midland & South Western Junction Railway (M&SWJR) was constructed by a series of companies, with the M&SWJR eventually operating a route from Cheltenham Lansdowne on the Midland Railway to Andover on the London and South Western Railway (L&SWR). Connecting two mainlines as a through route, the railway was a classic example of Victorian infilling of the railway system. The company had an interesting collection of locomotives constructed by several builders, including Beyer Peacock, Sharp Stuart and Dübs. Beyer Peacock 2-4-0 tank No.8 is seen on a passenger train at Cheltenham Lansdowne c.1900 in its distinctive maroon livery, with shaded lettering. The independent company always suffered from lack of finance, with this reflected in an eclectic fleet of locomotives, purchased when funds allowed. The line also featured light rails which had a tendency to break, a handicap which severely restricted the weight of the locomotives the Company could purchase.

One of three Dübs 2-4-0 tender locomotives No.10 constructed in 1894 for the M&SWJR is captured in lined maroon livery c.1920. They became GWR Class 1334 and were allocated the numbers 1334 to 1336 in 1923 and were later reboilered at Swindon Works. They proved useful locomotives, the only former M&SWJR design to enter British Railways Western Region stock in 1948, with the last example No.1336 withdrawn on 24 March 1954.

Beyer Peacock 0-6-0 tender goods No.28. There were two batches of these useful locomotives constructed for the M&SWJR, with the first batch of six constructed in 1899 and a further batch of four delivered in 1902. All of the ten class members were taken into Great Western ownership and reconstructed with Number 10 superheated taper boiler boilers, with this significantly altering their appearance. Allocated the GWR numbers 1003 to 1011, the last member of the Class 1005 was withdrawn in March 1938.

The Midland & Great Northern Joint Railway (M&GNR) was an amalgamation of several local railways in North Norfolk. Later under the umbrella of the Midland Railway and the Great Northern Railway, the M&GNR was incorporated as a joint operating company in 1893. Beyer Peacock constructed C class 4-4-0 tender locomotive No.34 for the Eastern & Midland Railway, one of the constituents that later made up the M&GNJR. No.34 is seen with a passenger train at Cromer Beach Station in 1888. They originally featured round top fireboxes but were later reconstructed with Midland Railway Johnson boilers.

S.W. Johnson designed 4-4-0 tender locomotive No.18 on an express service c.1905 and here it is seen in its distinctive lined yellow oaken livery. The M&GNJR's small works at Melton Constable could turn out an engine in immaculate condition. Copying both GCR and Beyer Peacock practice, the Company's engines sporting a central wheel to secure the smokebox door. These handsome 4-4-0 tender locomotives were the mainstay of the fast services from the East Midlands to the North Norfolk Coast until 1 October 1936, when the London and North Eastern Railway (LNER) took over the operation of the M&GNJR and then provided relatively more modern motive power to work services over the former joint railway. The locomotives were then given a 0 before their former M&GNR number to denote their place in the LNER's duplicate list of ageing locomotives. A small fleet of non-standard 4-4-0s more suited to Derby than Doncaster, the last was withdrawn in January 1945.

S.W. Johnson 4-4-0 tender express locomotive No.39 on a passenger service c.1900. These distinctive engines again featured Johnson's fine lines and were supplied by Derby Works. The extra 4-4-0s enabled the M&GNRJ to operate a much improved service of through fast trains over the joint line. Most of these locomotives were later modernised with Deeley boilers by Derby Works, which somewhat took away their original beautiful looks, but ensured the locomotives had a longer span of life.

A Lancashire Derbyshire & East Coast Railway (LD&ECR) Kitson constructed Markham D class 0-6-4 tank No.33. An unusual wheel formation in Great Britain, No.33 is seen when new in 1906/7 with an extended smokebox. The LD&ECR had opened in 1890 and was designed to compete with the GCR, GNR and Midland Railway for lucrative passenger and goods traffic crossing from the west to east coasts. In a period of railway company amalgamation, the M&GNJR became part of the Great Central Railway in 1907. The LD&ECR linked the original Manchester Sheffield & Lincolnshire Railway with the east coast and allowed the Great Central to serve the Manchester and Sheffield area, with the east coast ports of Immingham and Grimsby, allowing a fast connection with the continent of Europe. They became Class M1 under LNER classification and were used on heavy goods services. Unusually for a tank design, they were fitted with scoops for use over water troughs to extend their range of operation on the main line. The final member of the class was withdrawn on 4 July 1947.

North of England

The Lancashire & Yorkshire Railway (L&YR) was incorporated in 1847, originally from the Manchester & Leeds Railway. Over the next fifty years, the company greatly expanded over both counties, rivalling a number of other large northern railways, including the Midland, the London & North Western (L&NWR), the Great Northern and Great Central. The L&YR had a good working relationship with the L&NWR, who constructed some of their own designed locomotives for the former, much to the disgust of the Locomotive Builders' Association, who prevented it happening again. The company enjoyed a highly lucrative passenger and freight service. In 1912, it started to electrify suburban passenger services in Liverpool and Manchester and experimented with petrol railcars. They became part of the LMS in 1923. Locomotive 112 was an 11 Class 0-4-4 tank engine, designed by Barton Wright and built by Kitson & Co. of Leeds in 1873. It was reboilered in 1892 and withdrawn from service in June 1910.

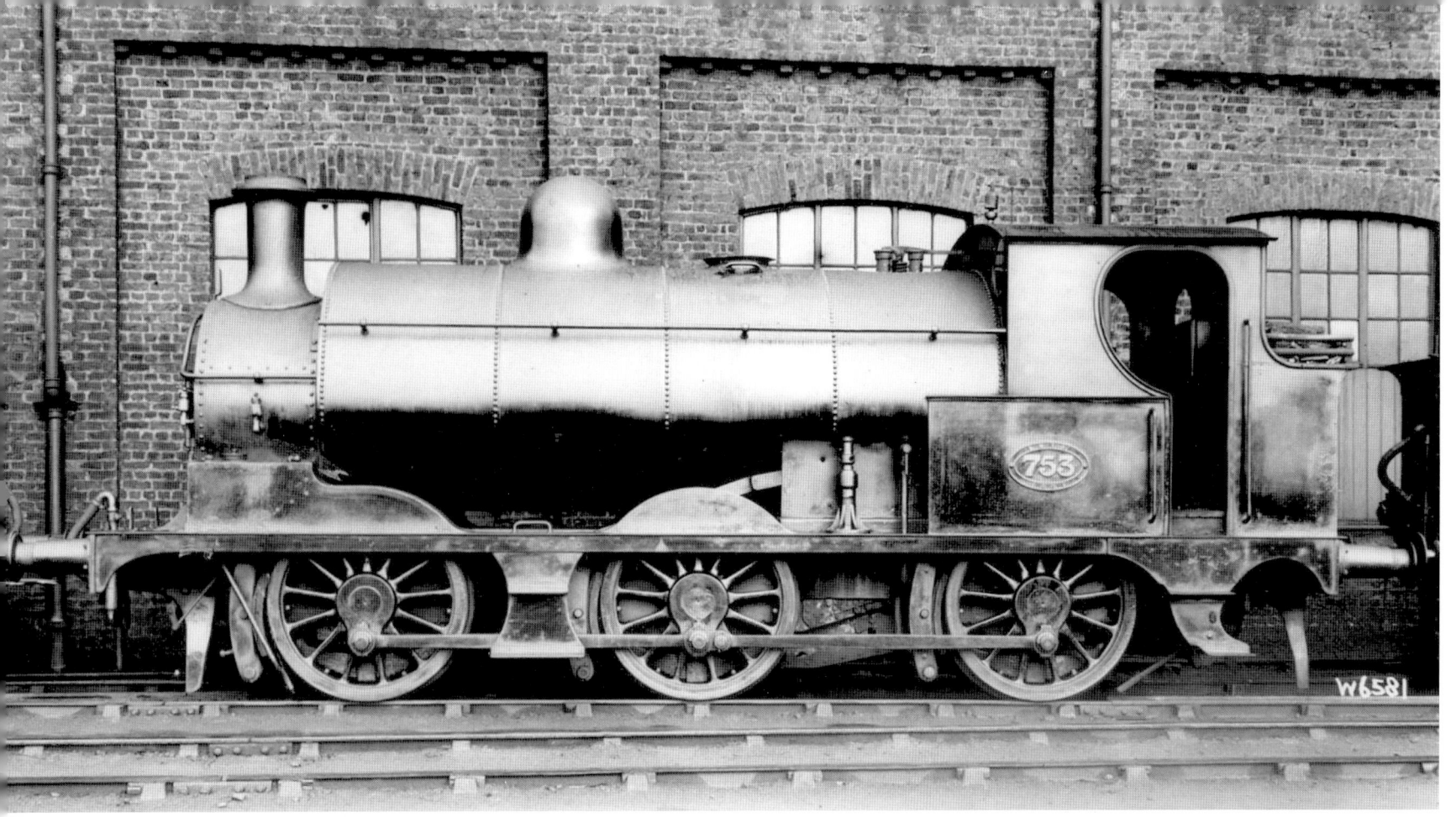

Originally introduced in 1876 as Class 25 0-6-0 tender goods locomotives under L&YR Locomotive Superintendent William Barton Wright at the Company's original works at Miles Platting; between 1891 and 1900 examples were reconstructed under Sir John Aspinall as Class 23 0-6-0 saddle tanks at the L&YR's new works at Horwich. The final surviving L&YR Class 23 saddle tank was withdrawn as a works' shunter at Horwich in December 1964. No.753 is pictured on shed c.1914. No.754 was sold by the LMS into colliery service in 1937 and became part of the National Coal Board fleet in 1947, with this enabling an otherwise extinct design L&YR design to survive in service into 1966 and subsequent preservation.

Class 25 Barton Wright 0-6-0 tender goods No.966 c.1906 is seen on shed in its ornate fully lined black livery, with shaded lettering. No.966 was one of a large group of goods locomotives which were the backbone of the L&YR locomotive fleet, many were later reconstructed to form the Class 23 0-6-0 saddle tanks. Constructed by a multitude of manufacturers, including Beyer Peacock, Kitson, Vulcan Foundry, in addition to the L&YR works at Miles Platting, Manchester, John Aspinall would continue production as his Class 27 at Horwich Works. First introduced in 1889, the class, as with a generation of previous L&YR and L&NWR designs, featured simple to assemble inside Joy valve gears, while the class would feature in experiments with superheating in 1912 under the Company's final CME George Hughes. The class would feature both round top and Belpaire fireboxes. Under the LMS, the long lived 0-6-0s were classified as 2F, with the last example withdrawn of an original Class 23 withdrawn by British Railways, London Midland Region in July 1959, while the last Aspinall Class 27 would survive in service until 12 December 1962.

Barton Wright 4-4-0 tender locomotive No.653 was constructed at the Atlas Works, Vulcan Foundry, Newton Le Willows in 1886. The 4-4-0 was one of a class of intermediate 4-4-0 passenger locomotives used on fast and semi fast train services on the L&YR. These versatile machines provided the Company with efficient reliable larger locomotives to haul its passenger services at a time when the trains were becoming heavier after the introduction of bogie carriage stock. The competitive service of trains to the Flyde Coast for Blackpool North, the joint L&NWR & L&YR Central Station and Fleetwood meant that Manchester's leading businessmen could travel in genuine luxury in their own club train to the coast twice a day from Manchester Victoria. Most were withdrawn by 1914, though two examples would survive to be classified as Class 2 under George Hughes in 1919. No.653 is captured in new condition with its original four wheeled tender.

Sir John Audley Frederick Aspinall became Locomotive Superintendent of the L&YR in 1886. He would go on to produce some of the company's most distinctive designs during his tenure of office. 4-4-0 express passenger locomotive No.1096 was introduced in 1891 as a new larger passenger type to supplement the increasingly outclassed Barton Wright passenger designs on principal express workings from Leeds Wellington Street, Manchester Victoria, Liverpool Exchange, Goole and Bradford Exchange. These handsome 4-4-0 tender locomotives were a significant step up in terms of haulage. The first batch of these locomotives was paired with tenders cascaded from the 568 class locomotives and were classified as type 3 by the L&Y Railway and later classified as 2P under the LMS.

L&YR Aspinall Class 91 0-8-0 tender heavy goods locomotive No.676 on shed c.1901. Constructed between 1900 to 1908, these powerful locomotives were designed for heavy long distance goods workings, especially long coal trains from the Lancashire and Yorkshire pits, a highly lucrative traffic for the Manchester based Company. The 91 class was first introduced in 1900 and featured two cylinders and Joy valve gear, with thirty-three members of the class reconstructed under George Hughes with larger boilers and re-classed as Class 30. The class were classified as 5F goods engines by the LMS and lasted into the early 1930s before the final withdrawals took place.

Two of the famous A.J. Aspinall Class 21 type 0-4-0 saddle tanks Nos.19 and 20 on shed in what is believed to be Liverpool where a large number of these celebrated 'Pug' locomotives worked the extensive series of lines linking the docks of the city. With two small outside diameter 13in x 18in cylinders, the Pugs were a derivative of three 0-4-0 saddle tanks originally supplied by Vulcan foundry in 1886. From 1891 to 1910, Horwich Works produced a further fifty-seven examples to a slightly modified design. They were classified as Class 21 under the L&YR's classification and 0F under the LMS. They were long lived, useful locomotives with a short wheelbase, perfect for the tight curves encountered in quays, with a number later being sold to industry given their diminutive overall weight of only 21 tons. The last example was withdrawn by British Railways London Midland Region in September 1964.

Furness Railway Class 120 (K1) 4-4-0 tender locomotive No.123 as constructed by Sharp Stewarts & Company c.1920. Designed under Furness Locomotive Superintendent William Frank Pettigrew, No.123 was one of three K1 class introduced in 1890 for passenger work. Their tenders featured outside springs, while the chimney was notably tall in appearance, with the class earning the title of 'Seagulls' with Furness footplate crews. The Furness Railway operated a busy system of lines throughout Cumberland. This included the Windermere branch to the Lakes, as well as the Cumbrian Coast Line running from Carnforth in Lancashire and a series of important lines serving the shipbuilding, iron ore and smelting industries centred on Barrow-in-Furness. The Company purchased a sizeable number of Sharp Stuart locomotives over the years and like the Welsh Cambrian Railways, the Sharp Stuart products lasted a long time in service, with many being taken over at grouping in 1923 by the LMS.

All three eventual Pettigrew 4-4-0 passenger designs bore a striking resemblance to other product of Sharp Stewarts, most notably the Cambrian Railway's Class 61 of 1901, with this begging the pertinent question of how much of the design of the various Furness 4-4-0s was actually independently developed under Pettigrew at the Company's Barrow-in-Furness Works. Though displaced on heavier passenger duties by Pettigrew's giant Baltic 4-6-4 tank design of 1920, the class of tidy 4-4-0s would still help well into the 1920s by supplementing the service to the Lakes on busy peak Summer Saturdays and on Bank holidays. All three members of the class survived past the 1923 grouping into the LMS, with No.123 as LMS No.10134 taken out of service in September 1927. Note the standard Sharp Stewart 0-6-0 for goods service in the background.

Mersey Railway Beyer Peacock Class 1 0-6-4 Tank No.1 *The Major* is seen on a passenger train of four wheeled carriages c.1901. Like their smaller Beyer Peacock 4-4-0 Tanks forebears on the Metropolitan and MDR, these locomotives were condenser fitted for working through the tunnel. With their thick outside frames, protruding outside cranks and bulbous condensing pipework, like Beyer Peacock's previous products for London's Underground, in no sense of the word could the small fleet of nine dating from 1885 be described as leading examples of elegance in late Victorian locomotion. However, the 0-6-4s worked well enough, though even with condensing gear fitted, conditions under the Mersey were at times almost unbearable given the dripping heat for footplate crews and constantly sulphurous breathing air for confined passengers in the packed compartments. No wonder and especially after the successful opening of the electric 'tube' City and South London Railway in November 1890, questions began to be asked as to when the Mersey Railway would benefit from similar electrification. The Mersey Railway opened on 20 January 1886 and linked Cheshire with Lancashire via a tunnel under the River Mersey, with six stations linking Liverpool Central to Rock Ferry. The line was electrified using 600v dc American Westinghouse Company equipment and rolling stock and re-opened on 3 May 1903. On electrification, No.4 *Gladstone* had been retained by the Mersey Railway until 1907 for departmental use, but was then replaced by recently withdrawn ex-Metropolitan Railway A Class 4-4-0 tank No.61. The railway was never absorbed in to the 'Big Four' in 1923 and remained independent until 1948 when it became part of the London Midland Region of British Railways.

Mersey Railway Beyer Peacock Class III 2-6-2 Tank No.13 on shed c1899. The ten locomotives consisted of two distinct classes; six Class IIs constructed in 1887 by Beyer Peacocks and three Class III produced by Kitsons and Company in 1892. Visually they were difficult to differentiate from each other, though both were fitted with the same trademark ugly condensing pipework to connect the smokebox with the side tanks, as well as large polished oval brass number plates positioned in the centre of the side tanks. They were later sold on after the line was electrified in 1903, with the entire class of six Class II and a single Class III purchased in 1913 by the Alexandra (Newport and South Wales) Docks and Railway (ADR) in South Wales to complement its four former Mersey Railway class Is. These became ADR No.6-11, while two Kitson Class 3s were sold to a colliery railway. When the ADR was taken over by the Great Western in 1923, all the former Mersey Railway locomotives became GWR stock, with the last not withdrawn until May 1932.

No.349 was one of a class of shunting locomotive introduced by Edward Fletcher for the North Eastern Railway in 1875. They were used in larger goods yards for general shunting and the marshalling of goods trains. Most of the class had been withdrawn by 1914, with a number being sold to industry, mostly colliery railways.

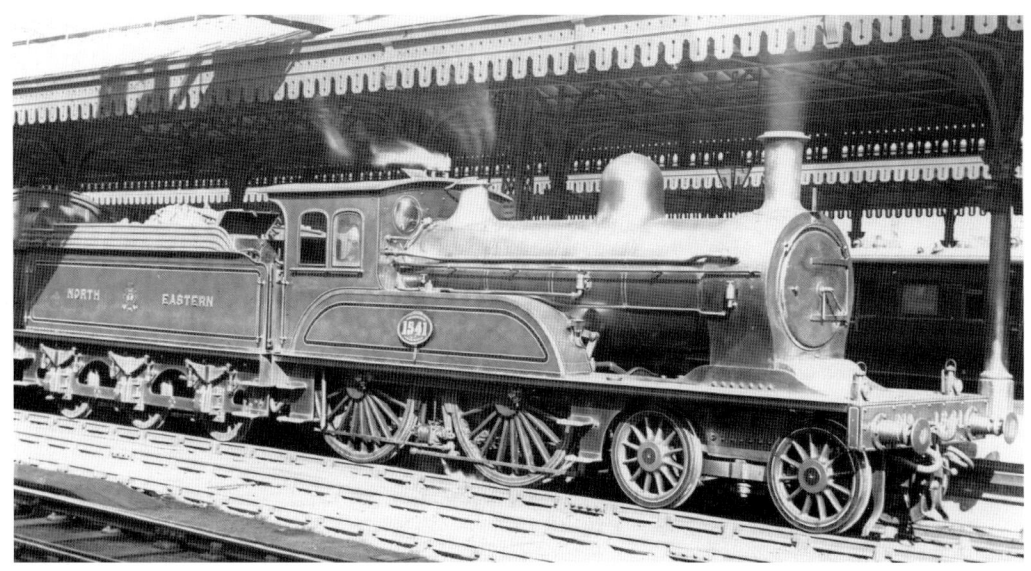

The development of North Eastern Railway locomotives is complicated and involved the gradual absorption into a greater NER of a string of companies, many of which such as the Stockton and Darlington and Newcastle and Carlisle could trace their ancestry back to the very earliest days of steam locomotion in this country. As a result, by the late 1880s, the NER had inherited a motley crew of diverse types, a non-standard fleet of austere primitive goods and especially passenger types. An unacceptable state of affairs and the appointment of Thomas W. Worsdell in 1885 and from 1890 his brother Wilson Worsdell would bring a much-needed degree of sanity to NER locomotive design, a timely change in leadership, especially as competition between the East Coast and West Coast group of companies was hotting up for the prestigious and lucrative Anglo-Scottish traffic. This would culminate in the famous 'Race to the North', with the NER expected to play its part in the renewed challenge between Euston and King's Cross to Aberdeen of 1895. Fortunately, the NER had recently introduced a highly capable 4-4-0 design in Wilson Wordsell's Class M1 of 1892. Known as 'Railcrushers' given the girth of the 160psi boiler, four 7ft 1¼in driving wheels and gross 50 ton weight, Worsdell would initially hedge his bets by producing a solo example of Class 'M' No.1619 in compound two cylinder form, with later in 1914 the engine modified to Smith three cylinder NER Class 3CC compound form. However, Wilson Worsdell decided in future to opt for purely simple expansion, with the younger Worsdell also producing a conventional Class 'M1' inside two cylinder design. With the rebuilding of the original compound 'M' into simple form, all later became NER Class M. Both the lone compound No.1619 and simple versions featured the same commodious glazed window cab for the crew. Throughout the 1895 race to Aberdeen, Class M1s were used to haul the 120 ton trains through to Edinburgh, Waverley from Newcastle without the need for double heading. No.1541 is seen c.1914.

NER Q Class 4-4-0 designed by Wilson Worsdell. No.1877 was introduced in 1896 for express passenger work around Hull, York and Newcastle and was a slightly lighter by two tons build on from the M1 class. Designated as Class D17/2 after 1923, both the M and Q classes lasted intact through most of the LNER period, with the final two members not withdrawn until just after nationalisation on 7 February 1948.

NER P3 0-6-0 tender goods locomotive No.2338. A robust, simple, classic British inside cylinder 0-6-0, the addition of a larger boiler by Wilson Worsdell would help produce a powerful and long lived design for heavy mineral work around the Northumberland coal field. A new NER Darlington Works and the amalgamated North British Company of Glasgow, as well as Beyer Peacocks of Manchester and Robert Stephensons of Newcastle; would all produce batches of the capable 0-6-0 until 1923. These versatile machines would become LNER class J27 in 1923, with the last member of the class withdrawn in September 1967.

Wilson Worsdell introduced his O Class 0-4-4 Tanks in 1894, with examples constructed between 1894 and 1901 for the NER. The locomotives featured on suburban and branch line work, with No.1438 pictured on a suburban working an unknown location c.1900 in its ornate original North Eastern green livery. The class became Class G5 under the LNER in 1923, with the last withdrawn under British Railways, North Eastern Region in December 1958.

Isle of Man

Isle of Man Railway Beyer Peacock 2-4-0 tank locomotive No.4 *Loch* constructed in 1874 with a small boiler. This was subsequently replaced in 1909 with a medium boiler. The famous 3ft narrow gauge Beyer Peacock 2-4-0 tanks have long been the principal form of motive power on the island. *Loch* was named after a former governor of the island and is still very much an active member of the Isle of Man Railway fleet of locomotives and is maintained at its highly capable Douglas Works.

Beyer Peacock 2-4-0 tank No.12 *Hutchinson* with a train of three bogie carriages at St Johns Station c.1923. No.12 was constructed in 1908 and was after the line was nationalised by the Manx Tynwald in 1978 was re-boilered with a larger boiler constructed by Hunslet in Leeds.

British Light Railways

West Sussex Railway, otherwise known as the Hundred of Manhood and Selsey Tramway owned three Manning Wardle contractors' type 0-6-0 saddle tank locomotives. One of these, *Sidlesham* 1861, constructed at Selsey c.1900 is seen in fully lined dark green livery. The line was constructed by the indomitable Colonel Holman Fred Stephens under the terms of the Light Railway Act of 1896 and the trio was purchased due to their light construction to run over the tramway's thin section rails, ash ballast and somewhat ramshackle infrastructure. The line ran from Chichester to Selsey in West Sussex and served a number of rural villages and hamlets along the route. In December 1910, the track was inundated by seawater when the embankment failed near Pagham Harbour. In 1921, Stephens would purchase a trial Wolseley-Siddeley petrol railcar and in 1924 two primitive railcars built on Ford Model T chassis, with bodies constructed by Edmunds of Thetford. *Sidlesham* was operated on the line until the tramway closed completely in January 1935.

The 3ft gauge Rye and Camber Tramway was opened in 1895 and ran initially from Rye to Golf Links Station that served the Rye Golf Club. It was later extended to Camber and became popular with day trippers heading to Camber Sands. The locomotives and rolling stock were supplied by W.G. Bagnall of Stafford, while later a locally constructed bogie carriage built by the Rother Iron Works was added to stock. One of the two Bagnall 2-4-0 tanks *Camber* is seen at Golf Links Station c.1925. The larger of the two was named *Victoria*, while later the Company acquired a small petrol tractor from the Kent Construction Company to work services. The tramway was taken over in 1939 by the Royal Navy and never reopened after the war.

The Rother Valley Light Railway opened in 1900 from Robertsbridge on the Tonbridge to Hastings line of the South Eastern Railway to its original terminus station at Tenterden (now Rolvenden). The line was later extended up the 1 in 50 incline to Tenterden Town Station and again extended in 1905 to Headcorn. The railway ordered two Hawthorne Leslie 2-4-0 tank locomotives built in 1899 for the opening, No.1 *Tenterden* and No.2 *Northiam*. In addition, four wheeled carriages were constructed by R.Y. Pickering. The line changed its name to the more familiar Kent & East Sussex Railway (K&ESR) in 1905. No.1 *Tenterden* is depicted just before opening and heading a train of original carriage stock at the first Tenterden terminus in 1900. Note the newly constructed locomotive shed in the background.

Rebuilt L&SWR Class 282 Ilfracombe 0-6-0 tender goods locomotive No.7 *Rother* at Rolvenden shed c.1908. The locomotive was one of eight directly ordered by Beyer Peacock of Manchester in 1873 by concerned members of the Board of the L&SWR following frustration at William Beattie's perceived lack of activity in delivering suitable motive power for use on the new Ilfracombe line in Devon on time. Examples were later reconstructed by William Adams with Class O2 type boilers and modified cabs. The L&SWR withdrew the class in 1913, though two examples were sold to the K&ESR in 1910 and 1914. Given their light weight of only 25 tons, they were perfect for the K&ESR's light rails and infrastructure. Colonel Stephens maintained an excellent working relationship with Dugald Drummond at Eastleigh Works and subsequently purchased another example of the class in 1911 to work on his equally ramshackle Shropshire and Montgomeryshire Railway No.3 *Hesperus*. The two K&ESR examples became No.7 *Rother* (L&SWR No.0349) and No.9 *Juno* (L&SWR No.0284), with both falling out of service after 1934 and being finally sold for scrap by the light railway in September 1940 as part of the wartime national scrap drive.

The Lynton & Barnstaple Railway (L&BR) opened in May 1898 and operated a 1ft 11½in narrow gauge line from Barnstaple Town Station on the L&SWR to Lynton in North Devon. The driving force behind the construction of what would soon become one of the nation's most cherished narrow gauge railways was Sir George Newnes, 1st Baronet, the wealthy publisher of the *Strand Magazine* and *Titbits*. Designed to cater for the increasing number of holiday makers and ramblers enjoying the splendours of North Devon, the line wove its way through delightful countryside. The small fleet of five tanks were all named after local rivers. The railway was taken over by the L&SWR in 1922 and grouped into the Southern Railway in 1923, who after investing in one new locomotive in the 1920s, closed the line on 29 September 1935. Manning Wardle 2-6-2 tank *Taw* was one of the four locomotives owned by the original company, is seen outside the shed at Pilton near Barnstaple at the time of the railway's opening.

The odd locomotive out in the L&BR fleet was 2-4-2 tank *Lyn* dating from 1898 and constructed by the Baldwin Locomotive Works of Philadelphia, Pennsylvania, USA. As with the mainline companies, a shortage of manufacturing capacity among British locomotive builders had forced the L&BR to look to America for additional motive power, especially as a prolonged strike by British workers fighting for an eight hour day in the industry had made the shortage of new locomotives even more acute. The sprightly 2-4-2 was a classic American outline design scaled down for British narrow gauge with three prominent domes on the boiler top, one in the centre a conventional steam dome, the other two were for sand. Bar frames rather than traditional British plate frames were used and it boasts an unmistakable American style 'railroad' headlight. On leaving Baldwin's plant in May 1898, the engine was immediately disassembled and shipped in crates across the Atlantic to the L&BR's small Pilton Works. *Lyn* is captured at Lynmouth Station sporting its second lined green livery, along with the engine's original copper capped chimney c.1922. A far less attractive stove pipe version would be later fitted.

Plymouth Devonport & South Western Junction Railway (PD&SWJR) Hawthorn Leslie constructed 0-6-2 tank *Earl of Mount Edgecombe* seen outside the locomotive shed at Calstock c.1910. Sister Locomotive *Earl St Levin* can be seen half inside the shed. Both 0-6-2s together with a 0-6-0 tank locomotive named *A.S. Harris* were supplied new by the builder in 1907 for the opening of the line from Bere Alston to Callington on 2 March 1908. *A.S. Harris,* as with the larger 0-6-2s, was named after a director of the Company and was a standard Hawthorne Leslie 0-6-0 tank with inclined outside 16in cylinders and box tanks. Another Colonel Stephen's enterprise constructed under the Light Railway Act of 1896, the PD&SWJR engines were originally painted in an attractive blue livery, with polished domes and brass work. All three were taken into the L&SWR in 1922 with *Earl of Mount Edgecombe* numbered as 757, *Earl St Levin* as No.758 and *A.S. Harris* as No.756 and were shortly after absorbed into the Southern Railway in 1923. Under British Railways Southern Region, 756, 757 and 758 became 30756, 30757 and 30758 in January 1948. No.30757 was the first to be withdrawn in April 1949, No.30756 followed in August 1951 No.30756. The 0-6-0 had long departed from the Plymouth area, working as a shed pilot and shunter and migrating right across the Southern Railway. Finally on 22 December 1957, 30758 was withdrawn having been in store at Eastleigh Works.

Liskeard & Caradon Railway (L&CR) long boiler and outside cylinder 0-6-0 saddle tank *Kilmar* constructed by Hopkins Gilkes of Middlesbrough in 1869 is pictured at Moorswater c.1906. The line was opened in 1844 to carry copper, tin and granite extracted from nearby quarries and mines immediately to the port of Looe in Cornwall. The railway was technically only a mineral route, though passengers were carried in open mineral wagons with benches fitted for the task, with the Company using a loophole to get around more stringent legislation on the conveyance of passengers. By the turn of the century, the small fleet of three L&CR locomotives looked to belong to a previous age, with *Kilmar* and *Cheesewring,* a very similar 0-6-0 saddle tank delivered from Hopkins Wilks' predecessor Gilkes and Wilson, originally only sporting primitive weatherboards for the crew. These were eventually replaced by covered cabs following heavy overhauls at Swindon Works along with a lone 2-4-0 *Lady Margaret* manufactured by Andrew Barclay and an occasional 'spot hire' experimental GWR 4-4-0 saddle tank No.13 working services; the L&CR was a most eccentric railway. In severe financial difficulties by the 1890s, the L&CR sought salvation by connecting with the equally financially insolvent Liskeard & Looe Railway, which then somewhat inevitably was taken over by the Great Western in 1909. Eventually the L&CR followed suit and also fell into the mighty GWR camp in 1917, with *Kilmar* becoming GWR 1312.

Swansea & Mumbles Railway large Hunslet 0-6-0 condensing tank locomotive was constructed in 1898. Two of a standard design ordered from the famous Leeds locomotive manufacturers, they become Mumbles Nos.4 and 5, bearing the brunt of passenger services until electrification of the tramway in 1928. Dating from 1804 and the earliest passenger railway in the world, the tramway was worked by horses until 1877 and was electrified at 650 V dc using double decker tramcars which could be operated in multiple as 'trains' from 1929. The tramway was later taken over by South Wales Transport Company, a bus operator at heart, which then not unnaturally went on to completely close the last section of historic line on 5 January 1960. One of the Hunslet 0-6-0 tanks is seen on a train of double deck and single deck tram cars c.1900.

The Wantage Tramway opened in 1875 and was one of Britain's best known steam roadside tramways. The two mile tram line linked Wantage Road Station on the GWR mainline from London to Bristol and South Wales to the important Oxfordshire market town of Wantage. Three of the line's locomotives are pictured from right to left. No.5 *Jane* was constructed by George England of New Cross in 1857 for £800, No.7 *Mary* constructed by Manning Wardle of Leeds in 1888 and an unnamed steam tram No.6. The latter was a most unusual totally enclosed 0-4-0 engine built to patented design by James Matthews in 1882. This featured a notably tall smoke stack chimney to help clear smuts and smoke away from the open saloons of the American style tramcar trailers used by the passengers. The tramway closed to passengers in 1925 and to goods traffic in 1945. No.5 *Jane* was purchased by the Great Western for preservation on the platform of Wantage Road Station as a permanent reminder of this characterful but sadly lost to time rural steam worked tram service.

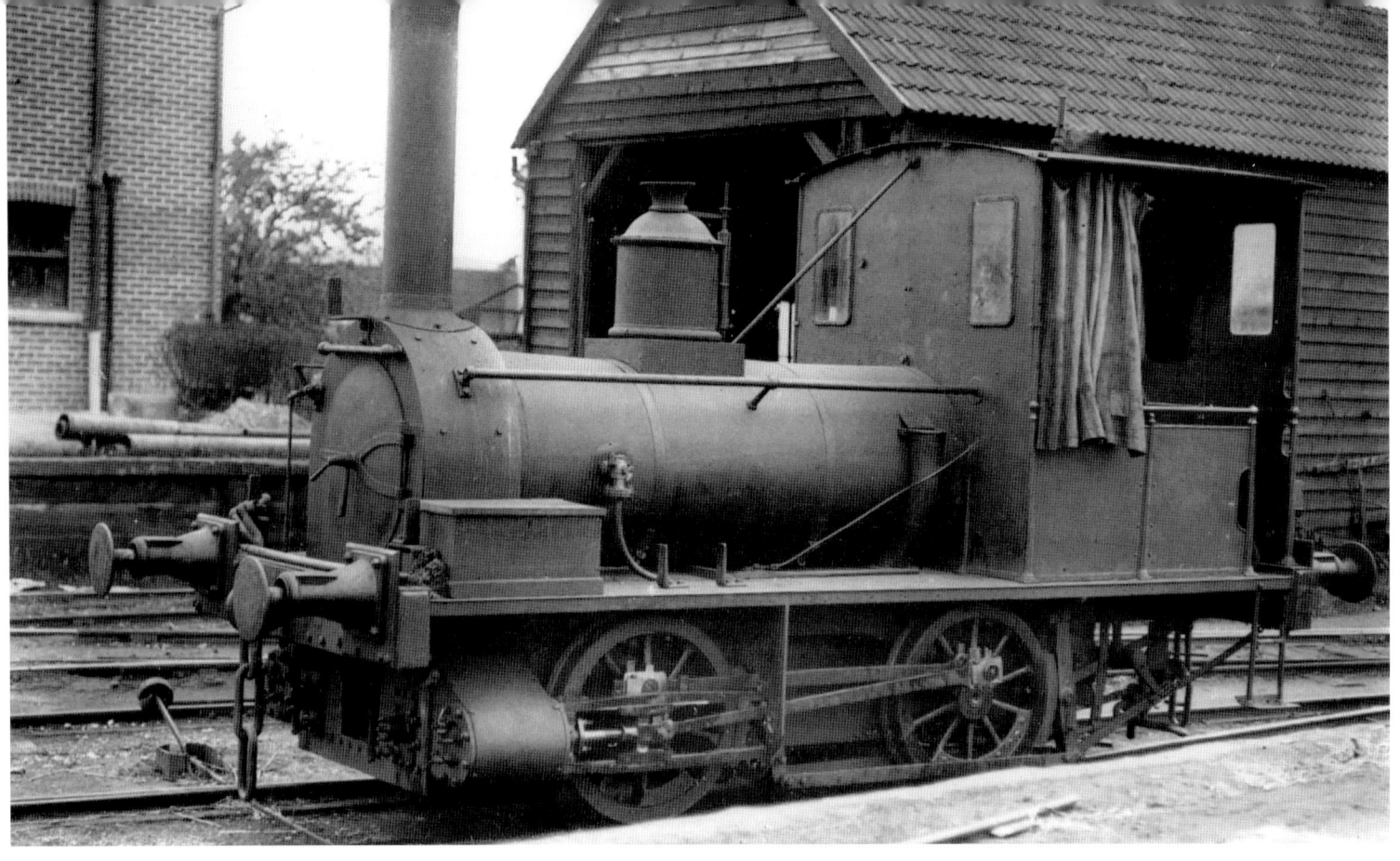

Wantage Tramway No.5 *Shannon* (*Jane*) at Wantage Town yard c.1920. This George England 0-4-0 well tank was constructed for the Sandy & Potton Railway in 1857 and then became No.114 in the L&NWR fleet which unsuccessfully trialled the 0-4-0 on the heavily graded Cromford & High Peak Railway without success. The engine would however find a successful niche as a Crewe Works shunter, renumbered No.1863. At the same time the little engine gained the name *Shannon*. In 1878, the engine was sold for £365 8s 1d to the Wantage Tramway, with the engine trundling under steam all the way from Crewe to Wantage via the West Coast Mainline and the GWR's Oxford route. In tramway hands, No.5 gained the unofficial name of '*Jane*'. An eventful life, prolonged by occasional visits to Swindon Works for heavy overhauls. The engine was finally withdrawn with the complete closure of the tramway in 1946. The GWR then purchased the engine for £100 from the receivers and then after putting No.5 through a well-earned spruce up at Swindon, put the engine on display at Wantage Station.

The Colne Valley and Helstead Railway (CV&HR) was opened in stages from 1861 to 1863 from Helstead to Chapel in the Wakes Colne, with a connection also made with the GER's busy Stour Valley line in 1865. This small concern continued to serve a largely rural community on the Suffolk and Essex border for exactly a century. The line closed to passenger traffic under British Railways on 30 December 1961 and to freight in April 1965, with the line demolished a year later. No.1 was a 0-4-2 tank constructed by Neilson in 1876 and Hawthorne Leslie 2-4-2 tank locomotive No.2, formally named *Colne* and dating from 1886, are both captured together at Haverhill c1906. In common with the GER, the Colne Valley and Helstead Railway was an air braked railway, hence the prominent Westinghouse pump fitted to locomotive No.1. No.1 was withdrawn by the LNER in 1923 and No.2 as LNER class F9 class was withdrawn in January 1930.

CV&HR Hudswell Clark 0-6-2 tank No.5 constructed in 1908.No.5 was purchased to haul good trains, which mostly consisted of brick traffic from Helstead brick works. The 0-6-2 tank became the single member of LNER class N18 and was renumbered as No.8314 after 1923.The engine was soon after transferred to Colchester for shunting and as non-standard was finally withdrawn in January 1928.

The Mid Suffolk Light Railway was opened in 1904 for goods and in 1908 to passenger traffic. The line ran from a junction with the GER at Haughley on the Liverpool Street to Norwich mainline to Laxfield in Suffolk. Constructed in the wake of the 1896 Light Railway Act, from the very beginning the railway struggled financially, failing to live up to expectations and only laying nineteen miles of route out of an ambitious intended original grand target of fifty route miles. After a protracted and complicated series of injunctions, bankruptcies and sackings, Hudswell Clarke were finally prepared to settle and release two engines to the struggling light railway. Hudswell Clark 0-6-0 tank No.1 *Haughey* had been constructed for the opening in 1904. It was followed by No.2 *Kenton* and finally by No.3 Laxfield. Though constantly short of funds, the three inside cylinder engines with their prominent Westinghouse pumps were kept in good condition in the reddish-brown livery and polished brass domes. Apart from No.1 *Haughey,* there is no evidence that the other two Hudswell Clark examples ever carried their original assigned names. The standards of maintenance of the permanent way left a great deal to be desired, with all three locomotives carrying jacks on the footplating due to the regular occurrence of derailments. With the Company in receivership for much of its independent life, all three locomotives eventually became LNER property in 1923. No.3 was immediately withdrawn, though No.1 and No.2 became LNER Class J64 and were given the new numbers of 8316 and 8317. Both locomotives were soon moved to Ipswich and Parkeston Quay for shunting, with No.8316 withdrawn in January 1928 and No.8317 in December 1929. The motley light railway would survive for a further unremunerative twenty-two years, finally succumbing to progress in July 1952 when British Railways closed the line to all traffic.

The Southwold Railway was a 3ft gauge line that connected Halesworth to Southwold along the marshy Suffolk Coast which first opened in 1879. No.1 *Southwold* was a 2-4-2 tank constructed in 1893 by Sharp Stuart of Glasgow as a replacement for the original No.1, a 2-4-0 tank which had to be returned to Sharp Stuart as the railway could not afford to pay for it on hire purchase terms, with the unwanted tank engine eventually sold to Santa Marta in Colombia. The second 'Sharpie' No.1 *Southwold* is captured at Southwold shed with the engine's crew c.1905. The Southwold Railway closed on 11 April 1929 and was left derelict until lifted and scrapped for the war effort in 1941.

The Leek and Manifold Light Railway opened in 1904 and ran from Waterhouses on the North Staffordshire Railway to Hulme End. It was designed primarily to serve quarries at Cauldon Lowe and Ecton creameries to send limestone and quality dairy produce respectively to Stoke, Manchester and London. The line had two Kitson constructed 2-6-4 tank locomotives No.1 *E.R. Calthrop* and No.2 *J.B. Earle*, with both engines named after the engineers involved in constructing the line. The railway was 2ft 6in gauge and had a positively Indian feel to it as the locomotives and rolling stock were designed by Everard Richard Calthrop who had previously made his reputation constructing lines for the British Raj. The line was similar in certain aspects to Calthrop's larger sub-continent Barsi Light Railway of 1897, not least in the use of transporter wagons to carry standard gauge rolling stock up the highly picturesque route. The first 2-6-4 tank design in Great Britain; they featured prominent colonial style headlamps, though they were never actually used in service, two outside cylinders, outside simplified Walshaerts valve gear and a fully enclosed glazed cab. No.1 *E.R. Calthrop* is captured in its attractive madder red livery and its large combined oval stock number and builder's plate and is seen hauling a train of primrose yellow bogie carriages at Hulme End Station in 1904. The line was grouped into the LMS in 1923 and closed to all traffic on 12 March 1934.

Bishops Castle Railway No.1 is photographed on a passenger train c.1930. No.1 was a former GWR Armstrong 0-4-2 tank No.567 constructed at Wolverhampton Works in 1869. The line was part of a projected railway from Craven Arms to Montgomery which fell victim to the Gurneys Bank Collapse in 1866. A wildly ambitious and doomed project from the start, the Bishops Castle Railway would open in 1866 from Craven Arms to Lydham Heath, with a branch to Bishops Castle. This nevertheless still left the line half finished and in the hands of receivers from 1867 to 1935 when it finally closed. No.1 was one of two principal locomotives used on the line in the latter years, with previously a hotch potch of small contractors' locomotives featuring, with this including two ex-Somerset and Dorset Joint Railway 2-4-0 tanks *Progress* and *Bishops Castle*, as well as another ex-GWR 0-4-2 tank *Perseverance*. Several attempts to sell the line to the GWR and the L&NWR failed as both companies understandably decided to steer well clear of an overly ambitious and highly speculative railway scheme.

Kitson constructed 0-6-0 tender contractor locomotive *Carlisle* is seen at Craven Arms station c.1908. Originally constructed in 1868 as a saddle tank and used in the construction of the Settle and Carlisle Line, the engine is seen attached to its original four wheeled tender. This was later replaced with an ex-GWR six wheeled Armstrong tender from Wolverhampton Works. Along with No.1, *Carlisle* was the railway's other principal engine used on services until complete closure of the route in April 1935. The engine had one last duty however, being used to help demolish the railway before being then finally cut up in 1936 at Craven Arms goods yard along with locomotive No.1.

Shropshire & Montgomeryshire Railway (S&MR) ex L&SWR Beattie Ilfracombe goods locomotive No.3 *Hesperus* at Shrewsbury Abbey Station shortly after the line reopened as a light railway c.1912. The S&MR originally opened in 1866 as the Potteries Shrewsbury and North Wales Railway and operated a line from Shrewsbury to Llanymynech on the Cambrian Railways, with a branch to Criggion to serve a quarry. By 1880 the line was already bankrupt, after which the railway was left derelict until Colonel H.F. Stephens promoted a light railway scheme in the 1900s. This led to the line reopening as a light railway for both passenger and goods traffic on 13 April 1913. As with much of the Colonel's light railway empire, the line made little profit and closed to passenger traffic on 6 November 1933. However, it would find a new role in 1941, being taken over by the War Department to serve newly built ammunition depots along its formation. Its new role would end on 29 February 1960 when the War Department and the quarry ceased to use the line.

The Easingwold Light Railway was privately owned throughout its existence and was one of Britain's very earliest minor railways, having opened before the Light Railway Act of 1896 in 1891. The railway ran from Alne on the North Eastern Railway to Easingwold for a distance of two miles. Hudswell Clark saddle tank No.2 is seen on a train of ex-North London Railway four wheeler carriages dating from 1872 at Easingwold c.1906. The 0-6-0 saddle tank locomotive was obtained in 1903 from Hudswell Clark and replaced the original 0-6-0 *Easingwold*, a very similar design previously constructed for the railway by the Leeds firm. The new No.2 lasted in service until 1947 when the Company started to hire locomotives from the LNER and later British Railways. North Eastern Region, with No.2 scrapped at Darlington Works in the winter of 1948/9. Always suffering the effects of road competition, the line closed to passengers on 29 November 1948 and to all goods traffic on 30 December 1957.

Scotland

One of four Caledonian Railway Class 88 2-2-2 tender passenger locomotive No.88 designed by Benjamin Conner and constructed at St Rollox Works, Glasgow by the Caledonian Railway in 1864. No.88 featured 7ft 2in single driving wheels and is pictured shortly after construction in a posed photograph with staff from the St Rollox and two four wheeled carriages.

Neilson constructed 0-4-4 Well Tank class 488 No.1167 of 1873. One of four ordered under Connor, the type would introduce the four wheel bogie truck to the Caledonian. Their outside low slung cylinders gave them a resemblance to the contemporary small locomotives designed by Matthias N. Forney for working the Elevated Railroads of New York.

Class 34 2-6-0 tender goods locomotive one of five converted from a Class 30 in 1912 by John K. McIntosh at St Rollox Works. The class were a development of McIntosh's successful class 812 0-6-0 design but with a front truck to help carry the additional 2½ ton weight of a Schmidt superheater. A less than convincing attempt to improve on the basic British 0-6-0, especially as both Churchward and Gresley were both designing thoroughly modern 2-6-0 equivalents at the same time, these locomotives were part of the fleet of five locomotives used on goods work on the Caledonian and were classified as 3F under the LMS. The last survivor was withdrawn as part of Stanier's radical standardisation policy in 1936.

Caledonian Railway 908 class 4-6-0 No.911 *Barochan* on a train of luggage vans at Glasgow Central Station c.1908. These impressive locomotives were designed under Caledonian Railway CME John Farquharson McIntosh. Introduced in 1906, the class featured smaller 5ft 9in driving wheels and 180psi boiler and were a mixed traffic development of McIntosh's famous class 903 *Cardean*, then a symbol of national pride in Scotland's railways and always seen in immaculate simmering Caledonian light blue livery at Glasgow Central every evening, with its unique ship style hooter booming down the Clyde a familiar sound for a generation of Glaswegians. Fitted with two inside cylinders and Stephenson valve gear, like the 903 class their hulking impressive looks were deceptive, with McIntosh like many of his contemporaries failing to make the leap from 4-4-0 to 4-6-0, this despite the fitting of Schmidt superheaters after 1911. The grouping and specifically the introduction of Henry Fowler's Royal Scot 4-6-0s in 1927, would see the somewhat disappointing and coal hungry class 908s taken off the heaviest and fastest passenger duties, with *Barochan* renumbered as LMS No.14612 and withdrawn in November 1931.

North British Railway 0-4-0 tender locomotive No.811 here seen posed with its crew and a shunter c.1898. No.811 was one of a number of class Y10 0-4-0 tender locomotives constructed under Thomas Wheatley after 1868 and used for yard shunting and trip working with goods trains. Later NBR pugs would forego the tender and instead run with a wooden truck for additional supplies of coal for trip work. These distinctive locomotives were subsequently rebuilt under Matthew Holmes in 1902 and later under William Reis in 1911 with new boilers. A useful and long lived class, examples would survive under the LNER until December 1925.

NBR No.1009 was a reconstructed 2-2-2 tender locomotive and was fitted with a Drummond boiler and featured a primeval-looking four wheeled tender and captured in full NBR bronze green livery c.1900, originally introduced under the NBR's first Locomotive Superintendent Robert Thornton and constructed by E.B. Wilson of Hunslet, Leeds in 1849 as a Crampton type locomotive. The locomotives originally worked the Company's principal passenger duties, but were later cascaded on to lesser services as trains became heavier in the 1890s and 1900s. No.1009 was rebuilt several times throughout its life and was only finally withdrawn from service in 1907.

NBR No.126 was a Thomas Wheatley inside cylinder 4-4-0 tender locomotive designed for crack express work between Edinburgh Waverley and Dundee. Wheatley would introduce the first inside cylinder 4-4-0 design, a classic British type that would go on to dominate on Scottish and many English railways before 1914. The opening of Sir Thomas Bouch's ill-fated Tay Bridge on 1 June 1878 would cut hours off the journey time between Edinburgh and Dundee, with a convoluted transit via Perth or a notoriously choppy crossing of the Tay Estuary by ferry previously endured by the passenger. This meant the NBR required additional express passenger engines. Unique Wheatley's compound 4-4-0 No.224 would be the unfortunate engine involved in the Tay Bridge Disaster of 28 December 1879, gaining the nickname 'The Dipper' in consequence. The last examples of Wheatley 4-4-0s class 224 and 420 designs were withdrawn by the NBR between 1919 and 1921.

An unidentified Reid J Class 4-4-0 is seen on a ballast train c.1920. These simple, but well constructed locomotives were designed under William S. Reid at the NBR's Cowlairs Works in Glasgow. Known as the 'Scott' class, members were named after characters from Walter Scott's popular Waverley novels. They became Class D29 and D30 under the LNER classification scheme after 1923. Members of Class J were modernised after 1911 with initially 18-element Schmidt superheaters and later with the 24-row Robinson version. Known as 'Superheated Scotts' they became LNER class D30. Many lasted well into British Railways Scottish Region days on passenger and goods services, with the last D29 withdrawn in November 1952 and with the last D30 succumbing on 25 June 1960.

One of the famous William Reid designed Atlantic Class 1 4-4-2 express passenger locomotives No.905 *Buccleuch* introduced in 1911. Named after the Scottish Duke of Buccleuch, No.905 is seen heading an express service c.1924, becoming LNER C 11 class in that year. Like their Caley contemporaries in the Class 903s, the Reid Atlantics were giants for the period, pulling heavy expresses on the East Coast mainline from Edinburgh Waverley to Aberdeen and over the Waverley Route to Carlisle. The fitting of superheaters would begin in 1915, with those fitted becoming NBR class H. Under the LNER they would become simple class C10 and superheated class C11 respectively, with No.905 renumbered as LNER 9905. Supplanted from premier East Coast duties by new Gresley A1 and later A3 Pacifics, No.9905 was withdrawn from Carlisle Canal shed in September 1937, with the last Reid Atlantic gone by November 1939. Hopes that an example of this totemic Scottish class would be saved for prosperity were dashed with the outbreak of the Second World War.

An NBR Pug type 0-4-0 saddle tank here posed for the camera in c.1900. These useful locomotives were used for light shunting in yards with tight radius curves and as pilot locomotives at sheds. As with other 0-4-0s of the period, they featured outside cylinders, Stephenson valve gear and slide valves. The Caledonian Railway would also introduce its own versions of the classic 'Pug' 0-4-0 saddle tank design in the Class 264s and Class 611s. This example was constructed in 1870 by Black, Hawthorne and lasted in service until 1912, with Neilson and Company also supplying very similar 0-4-0 saddle tanks to the railway in 1882. The NBR would go on to construct its own small fleet of similar 0-4-0 saddle tanks at Cowlairs Works, with these becoming Class G in 1913 and Class Y9 under the LNER.

Highland Railway David Jones Class D 4-4-0 tender locomotive No.94 *Strathtay* constructed by Neilson Reid in 1894. A member of the HR Strath class dating from 1892 and constructed by Neilson and Company, these small but robust, eight wheel express passenger locomotives were designed to climb the 1 in 70 gradients of the Highland mainline to Inverness on the Company's passenger and fast fitted goods trains. A capable 4-4-0, they featured the classic Jones trademarks of thick frames, two inclined outside cylinders based on the 'Allan-Crewe type', cabs with rounded corners and a louvered chimney to improve the draft. Classified under the LMS power classification scheme as 1P, the final members of the class were withdrawn in 1930.

David Jones Loch Class 4-4-0 tender locomotive No.70 *Loch Ashie* introduced in 1896 and. constructed by Dübs & Co for passenger and fitted goods work. As with previous Jones designs, the Lochs were based on his Class L 'Skye Bogie' or Duke class design of 1882. Nominally designed under his successor at Lochgorm Works Peter Drummond, such was the success of the basic Jones 4-4-0 design that in spite of the increasing trend to construct inside cylinder 4-4-0s, these still featured the outside type, though no longer steeply inclined as previously. A useful design, during the First World War given the extraordinary growth in traffic over the Highland mainline to Inverness, Kyle of Lochalsh and Wick and Thurso on the Far North line to supply the Grand Fleet at Scapa Flow as well as the many Army encampments and barracks dotted across the Scottish Highlands; Peter Drummond would add an additional batch in 1917. Classified as 2P under the LMS, the last example dating from 1896 was not withdrawn by British Railways Scottish Region until April 1950.

Originally constructed for the 3rd Duke of Sutherland's own personal branch line to his estate at Dunrobin Castle in 1870, the Kitson of Leeds constructed outside two cylinder 2-4-0 tank No.118 *Gordon Castle* was one of a miscellaneous group of locomotives owned by the Highland Railway. Inherited by the railway in 1895, the original Dunrobin was rebuilt with a larger boiler and cylinders at the Highland's principle locomotive works at Lochgorm in Inverness. Rebuilt and renamed as Highland Railway No.118 *Gordon Castle*, the resemblance to a modernised L&SWR Beattie Well Tank was marked. The engine was used on the Fochabers branch as well as on piloting duties. It was again renamed and renumbered to No.118A *Invergordon*. During the First World War the 2-4-0 was loaned to the Great North of Scotland Railway to shunt at Aberdeen Harbour, with eventually by 1923 No.118A dumped out of service near Culloden Moor Station on the Highland mainline.

Class V No.23 as constructed in 1903 by Peter Drummond from parts of scrapped locomotives at Lochgorm Works. Known as 'Scrap Tanks', these 0-6-0 tanks re-used the boilers and wheels of older engines, with three members of the class produced in this highly cost-effective manner. As with previous Highland designs, they featured outside cylinders and were used on shunting duties and for trip work. A somewhat brutish looking design, with rather large coupled driving wheels for a shunting engine; they were classified as 2F under the LMS. The final example was withdrawn in 1932.

Peter Drummond Class W 0-4-4 tank No.25 *Strathpeffer* constructed in 1905 at Lochgorm Works for branch line work. The class was classified as 0P under the LMS, with No.55053 the last former Highland Railway engine in British Railways Scottish Region service, being withdrawn from Inverness shed on 16 January 1957.

William Cowan designed Class C Great North of Scotland Railway (GNSR) 4-4-0 tender locomotive No.3 posed for the camera with its crew c.1912. Constructed by Stephenson & Co in 1890, these small 4-4-0 mixed traffic locomotives survived to become LNER Class D39 at the railway grouping in 1923. They were principally used on semi fast and fitted goods trains, with the last member of the class withdrawn in February 1927.

GNSR Class V 4-4-0 tender locomotive No.25 constructed by Neilson & Co in 1899 for use on principal passenger services. Designed by William Pickersgill at Inverurie Loco Works, these fine locomotives were constructed by Neilson Reid. The Company had originally ordered ten examples, only to discover it could only afford to pay for five, with the final five sold on to the Scottish & East Coast Railway (SECR). A fine investment, the SECR tried to buy the original GNSR examples, but was politely turned down by a thoroughly embarrassed Pickersgill. After the First World War Thomas Heywood would modernise the design with superheaters as Class F. Both variants became LNER Class D40 under the grouping of 1923 and survived into British Railways days. The very last D40 to be withdrawn was No.62277 *Gordon Highlander* which had survived longer than the rest of the class as it was used on steam excursions until June 1966. As for the SECR examples, the last G class 'Scotsman' was withdrawn by the Southern Railway in 1925.

No.87 was a Class 17 0-6-0 tender goods constructed by North British Locomotive Company in 1910 and designed by James Manson at the G&SWR's Kilmarnock Works. The class was absorbed by the LMS in 1923 and were all withdrawn as with nearly the entire fleet of former Glasgow & South Western Railway (G&SWR) engines under the Stanier 'cull' as non-standard. In keeping with this ruthless modernising policy, the last Class 17 was withdrawn in September 1936.

The G&SWR operated an extensive system of lines in the south west and Lowlands of Scotland centred on its impressive terminus at Glasgow St Enoch Station, a station with a grand overall roof on a similar epic scale to St Pancras in London, with both linked by a highly competitive service of joint Midland and G&SWR express passenger and overnight sleeping car trains. James Manson at Kilmarnock Works would introduce a graceful outside two cylinder 4-6-0 in 1903. Constructed by the North British Company of Glasgow as well as Kilmarnock, the G&SWR Class 381s were introduced the same year as the mighty No.903 *Cardean* on the 'Caley' and proved more than a match for the St Rollox 4-6-0 on competing services over the G&SWR's own route to Carlisle Citadel via Kilmarnock and Dumfries. The only G&SWR design to feature Belpaire fireboxes, the two 21in x 26in cylinders were fitted with piston valves which drove the middle coupled wheel, whereas on *Cardean* the inside cylinders drove the leading axle. A then modern cutting edge design, the engines featured a commodious double bogie tender as well as a Schmidt superheater, and as with previous G&SWR designs, Stephenson valve gear and plain slide valves. However, Manson would not completely understand the benefits of superheating by mistakenly viewing the device as an opportunity to reduce working boiler pressure from 180psi to only 160 psi, a false economy. No.386 is captured at Carlisle Citadel c.1914. All were absorbed into the LMS in 1923, with the last example withdrawn in November 1934.

Irish Railways

Great Southern & Western Railway (GS&WR) Alexander McDonnell designed class 21 2-4-0 tender locomotive No.23 dating from in 1870 and seen at Kingsbridge Station Dublin (now Dublin, Heuston Station) c.1900. McDonnel having been removed from Wolverton Works by Richard Moon had crossed the Irish Sea and had taken up residence at the GS&WR's Inchicore Works in Dublin. He found a pressing need for additional modern motive power to work the principal Cork expresses from Company's Kingsbridge terminus. Drawing heavily on a set of Beyer Peacock drawings, with the Manchester Company already heavily involved in the design of a 0-6-0 design in the long lived and successful class 101 (GSR J15), McDonnel would then sketch a then powerful, but light 2-4-0 with 5ft 3in coupled driving wheels. The similarities with John Ramsbottom's recently introduced Newton class for the L&NWR were marked. The class proved popular with footplate crews at Glanmire Road depot in Cork and at Inchicore, Dublin and were long lived. Given the class designation of G4 by the GSR in 1925, the last original member of the class was withdrawn in 1928. Note the double smoke box doors, a feature on many Irish locomotives and dating from the earliest decades of railways in the British Isles.

GS&WR class 90 0-6-0 combined tank and inspection carriage No.92 as constructed by Inchicore Works for shunting between 1875 and 1890. Built as required in two batches, the four members of the class were in effect primitive railmotors; featuring two inside 10in x 18in cylinders. The tiny locomotive is seen heading a typical Irish rural motley assortment of a full brake parcels van, cattle wagon and a carriage truck c.1900. They were all later reconstructed as 0-6-0 tanks and used on branch lines, or for light shunting at docks and goods depots. Long lived given their light weight at only 23 tons including the original carriage section, they were perfect for traversing poorly laid permanent way, with the first member of the class withdrawn by the GSR as class 30 in 1930 and the last under Córas Iompair Éireann (CIÉ) in 1959.

The sole member of its class and designed by Dubliner Richard Maunsell, Great Southern & Western Railway (GS&WR) Class 341 4-4-0 express tender locomotive No.341 *Sir William Goulding* as constructed in 1913 at their Inchicore Works, Dublin. Originally schemed under an ailing Robert Coey, in certain aspects *Sir William Goulding* would pre-empt the later work of Maunsell for both the SE&CR and as CME of a greater Southern Railway, not least in the use of superheating in the form of a Schmidt version in the case of the class 341, Walschaerts valve gear and a large diameter boiler. These would again feature on Maunsell's successful rebuild of former Wainwright 4-4-0s and his sparkling in performance Class V *Schools* 4-4-0's of 1930. The locomotives were constructed to run on the Dublin Cork main line hauling the company's most important trains, including the smartly worked Cork boat trains to Queenstown to connect with the behemoth liners heading for Southampton, Cherbourg and New York, notably White Star's surviving great liner RMS *Olympic*. After 1925 an amalgamated Great Southern Railway (GSR) would classify the 4-4-0 as class D1. A one off and therefore non-standard, No.341 was withdrawn in 1928.

GS&WR Steam railmotor No.1. Introduced in 1904, the single railmotor was similar to Drummond's own design for the L&SWR and was also not a success, proving under-powered and incapable of hauling a trailer for additional passengers. A one off failure, it seems to have been an attempt to modernise the Company's earlier combined coach and locomotive class 90 design. It was withdrawn in 1912, with the carriage portion reconstructed into a conventional bogie carriage No.1118, while the tiny vertical boiler engine, with its power bogie featuring distinctive disc style wheels; seems to have survived at Inchicore as a shunter.

Thomas R. Grierson designed Dublin & South Eastern Railway (D&ESR) 6ft 1in driving wheel 4-4-0 express tender locomotive No.57 *Rathnew* as constructed by Vulcan Foundry in 1896. The locomotive was a fortunate survivor of an otherwise ravaged D&ESR fleet as Ireland's railways were devastated between 1918 and 1923 in battles between firstly the IRA and British Forces and later between renegade IRA supporters and Irish Free State Forces. It was later reconstructed with a Belpaire firebox by the GSR in 1925 and was classified as class D9. Again renumbered as No.452, the 4-4-0 was finally withdrawn in 1933.

Waterford & Tramore Railway (W&TR) 2-2-2 Well Tank No.483 constructed in 1858 by William Fairbairn & Sons of Manchester at Tramore Terminus Station, County Waterford. The W&TR operated a short branch between Waterford and seaside resort of Tramore, with the line not physically connected to the rest of the Irish 5ft 3in railway system. The railway had a bucolic collection of locomotives and early carriage and wagon stock, which included three Fairburn well tanks and Bury designed ex- London & Birmingham 2-2-2 locomotive, reconstructed from a 2-2-0 tender locomotive and dating from 1837. The line later acquired some more modern 0-6-0 tanks hauling the trains after the amalgamation into the Great Southern Railway in 1925, with the line finally closing under CIÉ auspices on 31 January 1960.

The Dundalk Newry & Greenore Railway (DN&GR) was a branch in County Louth, which opened in 1867. Owned by the L&NWR, the DN&GR's rolling stock directly copied Crewe and Wolverton practise, but adapted to the 5ft 3in Irish standard gauge. Operating between the three towns and connecting with a ferry service for passengers and freight between Greenore and Holyhead, this unique enclave of Euston management would also use the Premier Line's famous rich livery of blackberry black for locomotives as well dark plum 'Carriage Lake' and 'Coach White' for its six wheel coaches. There was originally lucrative holiday traffic to the fine Greenore Golf Club, opened in 1896 as well as adjacent imposing and opulent L&NWR hotel. As the line crossed the border, the DN&GR was not absorbed into either the Irish Free State Great Southern Railway in 1925 or the Ulster Transport Authority in 1948. The railway had a standard fleet of six Crewe Webb constructed 0-6-0 'Special' saddle tanks, these complete with L&NWR cast iron 4 ft 8½in coupled wheels, with these passing into LMS ownership in 1923. A time capsule of late Victorian railway practice, the line became British Railways property in 1948 and closed on 31 January 1951. Crewe 0-6-0 saddle tank No.3 *Dundalk* is depicted on a train of L&NWR Wolverton Works constructed six wheeled carriages.

The West & South Clare Railway was a 3ft gauge line linking Ennis on the main 5ft 3in gauge line from Limerick to Kilkee and Kilrush via a junction at Moyasta. The line opened in stages between 1887 and 1892. It became part of the GSR in 1925 and was taken over by the CIÉ in 1945, after which an attempt was made to eliminate steam traction by bringing in diesel locomotives for freight and diesel railcars. This did not prevent the downward spiral of the line's fortunes and it was closed down in 1961, the last of the Irish narrow gauge lines. The photo shows 4-6-0 tank No.1, Kilrush constructed by Hunslet in 1912 shortly after coming into service. It remained in service until the mid-1950s.

The 3ft gauge Cavan & Leitrim Railway opened in 1887 and served the rural community of Counties Cavan and Leitrim in southern Ulster. On opening the narrow gauge the Company acquired eight Robert Stephenson constructed 4-4-0 tanks. They were named after the daughters of the Company's directors, with the one exception of a patriotic No.8 *Queen Victoria*. No.4 *Violet* is captured c.1905. These attractive robust locomotives were later joined by a single 0-6-4. Tank Nos.5-8 featured standard tramway fittings of skirting over the wheels, a cowcatcher and bell and headlamp. The 3ft gauge locomotives sported an attractive lined green livery. The conflict in Ireland from 1918 to 1923 would see the nameplate on No.8 *Queen Victoria* 'liberated' with the creation of the Irish Free State. All the locomotives were taken over by the Great Southern Railway in 1925 and by CIÉ in 1945. All were withdrawn after the line closed completely on 31 March 1959, another victim of cross border politics and road transport in the form of the lorry and bus.

The first sections of County Donegal Railway on opening in 1863 was originally 5ft 3in gauge, but was later converted to 3ft gauge. By 1914 the County Donegal Railway had grown to over 100 miles of line. Sharp Stuart, Atlas Foundry 1881 constructed 2-4-0 tank No.1 *Alice*, one of the earliest 3ft gauge locomotives acquired by the Company. The railway became part of a joint committee administered by the Northern Counties Railway, later part of the Midland Railway of England and the Great Northern Railway of Ireland. The LMS took over from the Midland Railway in 1923, with later the joint committee composed of representatives of the Great Northern Railway of Ireland and the Republic of Ireland's nationalised transport operator CIÉ after 1 January 1945. Criss crossing an ill-defined border, passengers and goods crossing the border were subject to two way Custom and Excise inspection, a burden that increasingly after 1921 made travelling by the narrow gauge County Donegal, an at times slow process, a gift to competing bus operators in other words. The Joint Committee from 1947 began to close its branches and substitute buses for passengers and lorries for freight traffic, with advent of the Ulster Transport Authority in 1949 further accelerating a closure programme that was completed in February 1960.

County Donegal Railway Nasmyth Wilson constructed Baltic 4-6-4 tank No.15 *Mourne* captured in works photographic grey livery as constructed in 1904. The development of 3ft narrow gauge railways in Ireland had been spurred on under the Conservative and Unionist government of Lord Salisbury who after 1885 sought to 'kill Home Rule with kindness' by encouraging rural development. For example, as part of this common marketing for dairy products was introduced, with Kerrygold butter a famous example. The construction of narrow gauge railways in poor areas of the country was seen as a critical component of the policy, with their construction further encouraged by the passing of the Light Railway Act of 1896. Named after rivers and loughs in the province of Ulster, in original form the Baltic tanks steamed poorly, with a multitude of firetubes badly affecting steaming. Eventually the top row of small tubes was removed, while the introduction of superheating greatly improved their usefulness. One of four specially constructed for the line, they were used on mixed traffic work, with No.15 withdrawn in 1952.

Black Hawthorne constructed 0-6-0 saddle tank No.3 *Lady Boyd* as supplied to contractors Butler & Fry in 1879. A typical tough, uncomplicated design fully in keeping with its harsh working environment, the 0-6-0 was later sold to the 3ft gauge Ballycastle Railway, which acquired three of the type for the commencement of services in October 1880. The line connected the southern terminus at Ballymoney on the Ulster Railway, later Great Northern Railway of Ireland, with Ballycastle. The route was primarily designed to enable passengers to connect with to the Giant's Causeway Tramway by making it possible for day trippers to travel from the GNR of I's Queen Victoria terminus in Belfast. Under the Midland's Northern County Committee a unique combination of Derby and distinctive Northern Irish railway practise would emerge in both 3ft narrow and Irish 5ft 3in broad gauge forms. The Company struggled financially, though the takeover of the Northern Counties Committee after 1 July 1903 by the Midland Railway seemed to offer a brighter future. In 1923 the Ballycastle Railway was passed to the LMS Northern Counties Committee. The venerable *Lady Boyd* was supplemented by two Kitson 4-4-2 tanks under Midland management in 1908. Later the LMS would supply new corridor coaches displaced from the recently closed Ballymena and Larne Railway. The line would yet again face another change in management under initially a nationalised British Transport Commission in January 1948, who then sold on the Northern Counties Commission to the new bus inclined Ulster Transport Authority, who then promptly closed the entire 17-mile 3ft gauge to all traffic in July 1950.

Beyer Peacock 1880 constructed 2-6-0 saddle tank No.109, formerly No.5 of the 3ft gauge Ballymena and Larne Railway. This most unusual early narrow gauge 2-6-0 saddle tank was used on mixed traffic work on the line and would survive to become first Midland Railway, then LMS Northern Counties Committee property. The Ballymena & Larne had opened in stages from 1877 and operated a line from the provincial town of Ballymena to busy Larne Harbour where services connected with ferries to the important Irish Sea port of Stranraer in Dumfries and Galloway in Scotland. Taken into LMS as part of the Northern Counties Committee, the NCC was later absorbed into the Ulster Transport Authority in 1949, who then promptly closed the branch in July 1950, with the UTA replacing the railway service with buses.

The Great Northern Railway of Ireland was formed in 1876 by a merger of the Irish North Western, the Northern Railway of Ireland and the Ulster Railway. It provided the main line route between Dublin and Belfast. The BT Class of 4-4-0 tank engines were introduced in 1885 and a total of thirteen were built, but they proved inadequate when heavier bogie carriages were introduced and all were scrapped in 1921. No.1 is seen here at Armagh in 1909.

Index

Aspinall, Sir John, 40, 41, 44-6

Conner, Benjamin, 78-80
Cowan, William, 93

Drummond, Peter, 89, 91-2
Dunbar, James, 17

Fairlie, Robert, 10
Fletcher, Edward, 51

Hosgood, J.H., 15
Hughes, George, 42

Ivatt, Henry, 11

Johnson, Samuel Waite, 37
Jones, David, 89

Locomotive manufacturers: Avonside, 23; Baldwin, 63; Bagnall, 30, 69; Beyer Peacock, 29, 32, 33, 35, 49, 56-7, 61, 103; Black Hawthorne, 87, 108; Cowlais, 85; Derby, 38; Dübs, 33-4, 112; Fletcher Jennings, 26; George England, 20-2, 67-8, 71, 77; Hawthorn Leslie, 60, 64; Hopkins Gilkes, 65; Horwich, 45; Hunslet, 66, 104; Inchicore, 98-9; Kilmarnock, 95; Kitson, 39-40, 50, 73, 90; Lochgorm, 91-2; Manning Wardle, 25, 58; Miles Platting, 41; Neilson, 69, 87-8, 94; Robert Stephenson' 17, 93, 105; St. Rolex, 78-81; Sharp Stuart, 18-9, 33, 47-8, 72; Vulcan Foundry, 43, 46, 101; William Fairbairn, 102; Wilson, E.B., 83; Wolverhampton. 75; Wolverton, 103

Manson, James, 95
Maunsell, Richard, 99
McDonnen, Alexander, 97
McIntosh, John F., 80-1

Penydarren tramway, 7
Pettigrew, William Frank, 47-8
Pickersgill, William, 94

Railway companies: Barry, 18; Ballycastle, 108; Ballymena & Larne, 109; Barry, 18; Bishops Castle, 108; Brecon & Merthyr Jcn., 17; Caledonian, 78-81;

Cambrian, 19; Cavan & Leitrim, 105; Colne Valley & Helstead, 69-70; Corris, 28; County Donegal, 106; Dublin & Drogheda, 10; Dublin & Kingston, 10; Dublin & South Eastern, 19; Dundalk, Newry & Greenore, 103; Easingwold, 77; Festiniog, 10, 20-4; Ffestiniog and Blaenau, 25; Furness, 47-8; Glasgow & South Western, 95-6; Glyn Valley, 32; Great Northern, 10; Great Northern Railway of Ireland, 110; Great Northern Scottish, 93-4; Great Western, 9; Highland, 88-92; Isle of Man, 56-7; Lancashire, Derbyshire & East Coast, 39; Lancashire & Yorkshire, 40-6; Leek & Manifold, 73; Liskeard & Caradon, 65; Liverpool & Manchester, 8-9; London & North Western, 10; Lynton & Barnstaple, 62-3; Mersey, 49-50; Mid Suffolk, 71; Middleton Colliery, 7; Midland & Great Northern Joint, 36-8; Midland & South Western Jcn., 33-5; North British, 82-7; North Eastern, 51-5; Plymouth, Devonport & South Western Jcn., 64; Rhymney, 14-6; Rother Valley, 60-1; Rye & Camber, 59; Shropshire & Montgomery, 76; Snowdon Mountain, 7, 31; Southwold, 72; Stockton & Darlington, 7-9; Taff Vale, 14; Talyllyn, 26-7; Ulster, 10; Wantage, 67-8; Waterford & Tramore, 102; Welshpool & Llanfair, 29; West & South Clare, 104; West Sussex, 58; Ulster, 10; Vale of Rheidol, 30

Spooner, James, 23
Stephens, Colonel, 58, 76
Stephenson, George, 7, 9
Stephenson, Robert, 8
Stirling, Patrick, 10

Thornton, Robert, 83
Trevithick, Richard, 7

Westinghouse, George, 12
Wheatley, Thomas, 82, 84
Wordsall, Richard, 82
Wordsall, Wilson, 52-5
Wright, Barton, 40-3